The Self-Defense App

Kevin Brady

This book is licensed for your personal enjoyment only. It is not intended to be re-sold or given away to other people. If you would like to share this book with another person, please purchase an additional copy for each reader. If you're reading this book and did not purchase it, or it was not purchased for your use only, then please return to your favorite retailer and purchase your own copy. Thank you for respecting the hard work of this author.

The Self-Defense App

1

Lance Fielding stood with watering eyes looking down at his longtime friend, Dexter Smith, lying peacefully in his casket. At just forty-five, Dexter was too young to be lying there. Yet, there he was….all poised and well-tailored to the "T" with his shiny bald head and Army suit on showing his rankings. He had only been retired out of the Army for six months. It was so ironic for him to have survived combat in Iraq and Afghanistan to come back home and get accidentally shot by a stray bullet falling out of the sky during the New Year's Eve celebration.

The church was packed with people with lots of folks having to sit in folded chairs brought in by church ushers.

After a few moments, Lance's feet shuffled his frame away from the casket as the procession continued moving, and before he knew it he was back in his seat in the church pews. As Dexter's friends and family continued parading around to get their view of his body, Lance sat thinking about good times he had with Dexter. The times they went to school together, games together, clubs together. The times when Dexter would show others that he was no pushover and got into fights defending himself or his close friends. His big biceps and triceps that he had developed over the years along with his handsome face gave him static electricity in regards to picking up women. But after he had met his wife, LaShonda, five years prior, he had become a settled down family man. After he retired from the Army, he had been working at Galactic Electronics in Research and Development beside Lance for three years. Dexter and Lance had an unbreakable friendship bond like none other.

Soon, everyone was outside watching the planes fly over, hearing the twenty-one gun salute, and hearing "Taps" play on the trumpet. Then, a U.S. flag was folded and presented to LaShonda. She sobbed as was expected, but fortunately, her parents were there to comfort her and her three kids. Lance walked over to her.

"LaShonda, if there's anything I can do for you, you let me know okay?"

"Okay," she said between whimpers.

They stood with a few moments of ambiguity in their midst. Lance couldn't just pull away after his comment. So he stood there beside her relishing the moment as the coffin moved slowly downward to its resting place. She wiped her eyes and looked at Lance with a forced smile on her face.

"Dexter was so excited with the new app you two were developing at your company."

"Yeah, he and I have been putting a lot of time and energy into that app. Burnin' a lot of midnight oil. It's gonna be hard to continue developing it without him."

"Well, he would want you to continue without delay, Lance."

The casket stopped at its resting place. The gravediggers began shoveling dirt atop of it. It began to set in that Dexter Smith was no longer. People constantly made their way over to LaShonda to console her and would continue either on towards either their vehicle or to the church kitchen for repast. Lance kept standing and watching the dirt get tossed on his friend's casket until the very last dirt was tossed. He walked to the kitchen and got a plate to go and trekked to his car.

    Within minutes he was on the highway headed home. He began brainstorming about Dexter and their new app creation. It was all finished except for an image or person's likeness. With all the lawsuits amuck in the world, he couldn't just choose ANYONE'S image to use for their design. Then, it struck him. It was a perfect fit! They would use Dexter Smith's body dimensions and face in the app since he co-created the idea for it. It would be a perfect tribute to his late friend for all of his time and energy with the project. He smiled and reached for his CD player and started playing "Papa's Got a Brand New Bag."

***

2

Peter gave his mom a quick kiss and ran out the door with his book bag. "Love you," she hollered after him and he replied the same. He jumped in the Land Rover with Lance who awaited him with the vehicle running. They then drove off.

Lance looked over at him and asked, "Son, remember what I told you to do if you ever need help with a bully or a stranger, right?

"Fight the bully back and run from the stranger?" Peter answered looking at his dad for assurance that his answer was correct.

"That's right. You got it. But I got something ELSE for you too!"

"What do you mean?" Peter responded.

His dad handed him a brand new cell phone with a sticky note attached to it. "Wow!" Peter said full of excitement. "Is this for me?"

"Yes, son, it's for you. The phone number is on the sticky note. And see the big "H" button at the base of the phone under the little cage thingy?"

"Yeah."

"That's a new app I came up with. You open the little cage and mash that "H" button only if you need help in defending yourself at any given time. H is for help."

"Neat! What happens when I mash it?" he quizzed with his finger about to press it.

"NO…NO….don't mash it now! You'll see at the right time, " his dad insisted. "Just don't use it for nonsense, got me? It costs $19.99 every time the button is used, so make sure you use it for EMERGENCIES ONLY, follow me?"

"Okay. Thanks dad," he answered still smiling as he slid the phone in the pocket of his jeans.

"I mean it now! Only mash the button if you're in danger! Okay? I'll know whenever you mash it!" Lance hollered after Peter as he dashed up the school steps.

Peter walked briskly to the front entrance of his elementary school with his brown hair blowing in the morning wind. Even though he was just in the fourth grade, he knew how to work a smart phone like a grownup. His dad was a technology guru who worked in research and development for Galactic Electronics, a major cell phone maker, which enabled his family to keep the latest phones with the latest technology. This was the very reason that Peter could manipulate a smart phone better than most kids his age. He scooted to his class, plopped down in his seat, and slid the phone out of his jeans to check it out in detail. He could hardly wait for the time to present itself where he would need to depress this mysterious "H" button on his phone. He started going through the different applications for a few moments. A few of his buddies came gazing at him as he toyed with the phone and started asking him questions about it. After twenty minutes the bell rang, and it was time to get seated and listen to the morning announcements over the intercom. Peter slid the phone back into his jean pocket and got into school mode.   His science teacher, Mrs. Barbara Owens, took roll and began to get into the day's lesson.

   The day was moving right along. As Peter listened to his Mrs. Owens discuss photosynthesis, he got the urge to go to the restroom to pee. He held it so he could listen to all of the details about how chlorophyll in plants used sunlight energy to change carbon dioxide to oxygen.   After about thirty minutes discussing it, she moved on to the water cycle. It seemed as though talking about water made the urge to pee stronger and stronger.

   "So, this is how it works," she continued as she picked up a big poster board that had an arrow in the center. She started rolling the arrow around on the poster and resumed talking. "The sun causes water to evaporate from the ocean, and it makes the puffy white clouds become dark gray and full of water. Then……." Peter couldn't wait another minute. He went to the teacher and asked for the hall pass to go to the restroom. After she handed it to him, he stormed out of the classroom walking briskly down the hallway. He got to the first urinal he could get to and unzipped in lightning speed and began relieving himself. He heard a group of older boys come into the bathroom. It was four of them, and they were horse playing around with one another and laughing. They walked over to Peter and one used the other urinal beside him while the other three stood behind him grinning. He peeped around at them feeling a little suspicious of them. He got finished, quickly zipped up, and attempted to head away from the urinal but one boy put up his arm blocking him. "A….A….Not so fast!" he blurted.

"What do you want?" Peter asked.

"I want your lunch money! Now GIVE it to me, punk!"

"Yeah!" hollered the other three boys. "Or we'll smash your face!"

Peter reached into his pocket and gave them the five dollar bill he had.

" What's in the OTHER pocket?" asked one bully.

"Nothin'," Peter mumbled thinking about the new cell phone inside it.

"Let me SEE!" he retorted with a sly grin.

Peter thought about the button his dad had warned him not to mash. He then reached into the pocket and subtly felt for the little cage. He quickly reached beneath the cage, mashed the "H" button, and proceeded to slide the phone out of his pocket.

"Ohhhhh, look what we have HERE!" said the bully. "Now HAND it over before I break your nose!"

Just as he was about to hand it to him, a thick white fog instantly appeared in the bathroom. Just as quick as it appeared, it then vanished. Surprisingly, there was now a well-built black man standing near them with big biceps and triceps and rippling pecs topped off with a shiny bald head. The man frowned and stepped to the boys chopping at them, "If yall don't leave this boy alone, I'm gonna beat yall little asses!" The boys all froze in terror wondering where had this dude come from. The strange black guy snatched the phone from the bully holding it and handed it back to Peter. He grabbed the bully by the neck with his big opened hand and snarled at him , "Now get to steppin' before I put my foot in yall lil' asses! And don't EVER let me hear of yall messing with this kid again! Ya got me?" He shoved the boy by his neck in the direction of the door. All four boys took off running and sliding out of the bathroom. Peter kept gazing at the black man with his jaw dropped and was so startled by the magical appearance of the black stranger that a part of him wanted to take off running behind the bullies. He couldn't believe what had just happened. The guy reached over and pressed the "H" button on the phone looking into Peter's eyes and vanished once more. Peter was alone in the bathroom once again. He jammed the phone in his pocket and flew out of the bathroom back to his classroom.

Hysterically, Peter blasted to Mrs. Owens what had occurred. She was spellbound and glad he had had a good outcome with the older boys in the restroom. His description of the guy who came to his rescue really interested her though. The words, "black", "bald", "bare chested", "muscled up". She asked Peter if she could mash the button in the classroom to show the class, and told him that she would pay the twenty bucks if what he was describing was true. He allowed her to. The smoke came and vanished, and Mr. Muscles appeared just as Peter had described. His shiny, brown, bald head snatched from side to side searching for the sign of an altercation in the classroom, but could find nothing. He stood there with a shiny looking perfect muscular physique in nothing but red Speedos looking as if he was carved out of Hershey's chocolate. He then asked while cracking the knuckles of his fists, "Is there a problem?"

"Not in the classroom, there isn't," Mrs. Owens answered with a smirk on her face and a raised eyebrow. With her right hand on her hip and her left palm caressing her chest, she waited for his reply.

"Then, where IS the problem?" he asked looking puzzled with his chest pecs seesawing in his chest.

Mrs. Owens seductively pranced over to him and whispered something in his ear. Then she looked deeply into his eyes, grabbed his head from behind with both hands, and planted a kiss on his lips.

"Ooooooooh, Mrs. Owens!" the class teased.

"Yall do yall work!" she snapped at the class coyishly fanning her hand in the kids' direction and turned back to Mr. Muscles wearing a smile.

Mr. Muscles gave her a big smile and then licked his lips. "So, uh, what are you doin' later on?" he asked showing his snow white teeth. Mrs. Owens felt like she had hit the lottery. Her biological clock was ticking, and it was telling on her. At forty-five years of age and three years divorced, she was subconsciously allowing her primal instincts to come full circle. She grabbed him by his hands and began leading him to her desk on the far side of the classroom, but he disappeared into the thin air once they arrived at her desk. "Aw, fuck!" Mrs. Owens hollered before she could help it and cupped her hand over her mouth. "Yall didn't hear that!" she quickly followed. Then added, "Looks like I'm gon have to get ME one of THEM phones! Come here, Peter! Where yall bought that phone at?" she snapped with a frown.

When the bell rang at 3:00, Peter couldn't remember another word the teacher had said regarding his classwork. He stormed outside to wait for his mom to pick him up. By now, whispering had ensued and news of what had happened was spreading like wildfire. The bullies had spread a rumor around school that some muscled up bald headed black man was roaming around the school campus intimidating students for no reason. Peter knew better. Mr. Muscles did not show up until he mashed the magic button on his phone.

His mom pulled up on the curve, and he jumped in the car full of excitement. "MOM...MOM!! Guess what?" he hollered.

"What Peter?" she asked.

"You won't believe what happened today!"

"What happened?" she asked rubbing him on the head with a smile.

"Well, dad gave me this new phone this morning...." He pulled the phone from his jeans. "and see this button? He told me to mash it ONLY if I needed help, and I NEEDED help when four boys tried to rob me of my lunch money in the school bathroom. So, I mashed the button when one of the boys made me pull this phone out my pocket, and when I did a mean black guy with big muscles appeared and ran the boys away!"

"Peter, are you sure you didn't dream this?"

"No, this really happened!" he assured.

"Can I see the phone?" she asked. He handed it to her, and she examined it on quick intervals as she diverted her eyes back to the windshield to drive. "You mean, you mashed THIS button?" she asked. Peter tried to warn her not to but before he could stop her, she had already done so. In an instant, some white smoke appeared in the backseat of the minivan causing Peter's mom to panic. Peter told her that it was the same smoke he had seen. Then, just as quick as it appeared, it disappeared without a trace leaving a muscled up, shirtless, bald-headed black guy in the back seat. Peter's mom screamed. She snatched the van to the side of the road and hollered, "Run Peter Run!" Peter leaped out of the van and ran to her in front of it shouting, "Mom…Mom! That is the same guy who helped me out!" She was anxious to continue getting them to safety but slowed her desperation to listen to his reasoning. The guy stayed in the backseat of the car. His mom stopped running and listened to him until he got finished explaining. Then she cautiously eased back to the van shouting, "Get out of my van!   Who are you?!"

   The muscular black intruder opened his door, and she made a break to take off running again grabbing Peter by his arm in the process. He demanded, "Look lady……I'm not going to hurt you! I'm here to help you! Apparently, you need some self-defense help since you mashed the button!"

   "Who are you? How did you do that?" she asked in terror.

"Your guess is as good as mine, mam! So, you're telling me you don't need help?"

"No!", she hollered with a serious look in her eyes.

"Well, mash the damn button again, and my ass will be GONE!" he chopped.

"Why can't you just go wherever the hell you came from ON YOUR OWN?" she snapped.

"Cause that's not the way it works! I come and go only by teleportation upon the commands of phone owners who have the new Fielding App. Once I'm summoned by the button, I come and take care of business helping whoever's in danger however long it takes and then I disappear after the danger is erased or if there is no danger, I disappear in twenty minutes or sooner with the press of the button again. The button on your son's phone is a portal to my metaphysical world."

"You've GOT to be kidding me!" she chuckled. "The Fielding App?" she quizzed. She sighed in disgust and rushed inside the van to grab the phone. She flipped the cage open and pushed the button again really hard and tossed the phone back on the seat saying, "There!" The guy instantly disappeared without a trace however his voice uttered a loud "THANK YOU!" in his absence.

Peter's mom gave the command to re-enter the van. Once they were back inside, she then started it, jerked the lever to drive, and started down the road with frantic talking.

"Lance has got to give me some answers when we get home!," she said with a pissed off look on her face. Her blonde head jerked from side to side maneuvering in and out of lanes as she navigated back toward their house. The van couldn't get home fast enough, it seemed. "What in the world is this…..some kind of a game he's playing?!" she rationalized with a puzzled look on her face breaking the silence. She stroked Peter's hair and turned her attention back to driving the minivan. No sooner than she spoke, Peter's phone chimed with a message. He looked on the screen and seen a notification that $59.97 was now due which would be charged to the cell phone bill. "What was that?" she asked.

"A message saying we owe a total of $59.97 for the new 'Self-Defense App' fee," he said. "WHAT?!," she snapped. Then he added, "Dad told me this morning not to mash the button except for extreme emergencies since it would cost $19.99 each time it was used."

"So, this was the third time just now?" she asked.

"Yeah, my teacher, Mrs. Owens, did it the second time. She wanted to see how it worked once I told her about the boys attempting to rob me of my lunch money. She said she'd pay the $19.99 for THAT time."

His mom made a loud "AAARRRRRRGGGGGHHHHH!!!!" sound like she was fuming mad and was wearing a snarled look. They proceeded on down the road with his mom swerving around cars and driving frantically like he had never seen before. Part of him was excited and thrilled by the ride and part of him was scared they might get into a wreck before long. Within minutes they approached their street, and his mom made the turn making the tires scream in the process. Their van rolled into the driveway behind his dad's Land Rover, and his mom slammed it into gear. They both jumped out and headed for the front door. Once inside, his mom called out, "Lance…..Lance…..Where are you?" His dad came walking down the stairs wearing a smile. He was still wearing his shirt and tie from the office. His routine of working half days on Wednesdays allowed him to have been home since noon. "I already know what you're so perturbed about," he said with a grin.

"Oh, do you? Well, let's hear it!" she blasted.

"You wanna know about this new phone app, right?"

"YES!"

"Well, it's a new app I've been working on in my lab out back for the last several months, and I was letting Peter be the first to test out."

"Why didn't you tell me about all of this?" she quizzed.

"Honey, I wanted it to be a surprise is all."

"How in the heck do you make the black guy pop up when the button is pressed? That guy scared the Bejesus out of me!" she snapped.

"Was that cool or WHAT?" Lance asked. Then he added, "The dude you seen was my old friend Dexter Smith that just passed away, remember? I decided to download his picture into my computer and insert his dimensions so he would be the guy that shows up since he was so instrumental in helping me launch this thing!"

Lance wiggled his fingers in his extended right hand for Peter to give him the phone. When Peter handed it over, he began fiddling with it. "So, I see between you both the button was mashed twice," Lance commented as he fumbled with the phone's settings. "Did you need help in school, Peter, or were you goofing off when you pushed the button the other time?"

"No, these four guys pinned me up in the bathroom and tried to take my lunch money and my phone, and I secretly mashed the secret button as I pulled the phone out my pocket."

"Very good! And the black guy appeared and helpled you out?"

"Oh yeah! He offered to whoop their ass and chased them out of the restroom!"

"Watch your language, Peter!" his mom hollered.

"Perfect!" Lance shouted clapping his hand in satisfaction. He sat down on the sofa and cut the T.V. on with the remote while Peter's mom looked on still irritated with envy. He added, "See, I knew it would work, but I wanted to have it tested in a normal everyday type of way in a normal everyday setting. Now, what's for dinner, hun?"

"I don't know, I was thinking I'd let it be a SURPRISE since I had a big SURPRISE today from your phone!" she answered with sarcasm.

"Lisa, there's no need to be so sarcastic! You weren't harmed in any way, were you? So just drop it, will ya! I'm sorry I didn't tell you about it. I was going to eventually, but I didn't think Peter would have to use the app so fast. I'll make it up to you, okay?"

"With money?" she asked winking her eye at Peter.

"Alright, you win! With money!" Lance agreed. "How much are we talkin'"?" he asked curiously in a brace-yourself manner while looking over his glasses.

"Two hundred bucks"

"Okay, you got the two hundred dollars," he said dryly looking up at the ceiling.

"Let's order pizza!" Peter suggested.

"Sounds good to me," Lance stated.

"I second that notion," Lisa added. She ordered one on her cell phone, and then walked over and asked Peter for his cell phone. After he handed it to her, she sat down on the sofa beside him and began tinkering with it. She wanted to right beside him in case she had a question about something. The pizza showed up, and they all dug in. Lisa continued berating Lance with questions as she chewed.

"So, how did you get the defender to appear by the mash of a button?"

"If I tell ya, I'll have to kill ya," Lance teased looking at her with a grin.

"Tell me or I'll kill YOU!" she snapped back looking at him with her eyes stretched still chewing pizza.

"The defender appears by electromagnetic rays emitted by the phone," he explained. "It involves a bunch of physics stuff. Stuff like gamma rays, wavelengths, relativity, time, space, a whole bunch of formulas like e=mc squared....stuff like that. Understand?"

"Not really, but you can leave it at that," she said as she sat up on the sofa and tossed her crust back in the pizza box.

"Oh, now the kill you part," Lance added as he reached at Lisa with his hands raised closing in on her neck. He put them down before he got to her neck and instead reached at another slice of pizza.

"So, when are you going public with this?"

"Tomorrow at our board meeting," Lance began. "I'm planning to do a little show and tell for our board members. They're gonna be shocked to see Dexter Smith show up in the board room again with the click of the button."

"Everybody's gonna go crazy over this app! I actually think it's incredible, honey!" Lisa stated nodding her head having come full circle after having her questions answered, her steam blown off, and her stomach filled.

"So you forgive me?" Lance asked.

"Yeah, I guess. But, please, no more secrets, hun." Lisa said looking back over her shoulder as she headed down the hall to the kids' rooms.

***

3

The next morning, Lance walked into his office, put down his briefcase, and seated himself at his desk. He paged his secretary to come into his office. A couple minutes later she came to the doorway.

"What can I do for you, Mr. Fielding?" she asked.

"Uhhhh, can you get me a cup of coffee, please? Two creams….six sugars. As he waited for the coffee to be brought back, he began a mental play by play of his presentation. He pulled out his phone and checked his emails. His secretary stepped into the office with the cup of coffee and handed it to him. "Thanks, Mandy," he offered. "No problem," she replied and headed back out. Lance intercepted her departure with more conversation.

"Oh, I've been meaning to ask you…..Are things okay with you and Jack now?"

She paused and then looked down as if searching for the answer on the floor. Then she looked up, smiled, and said, "He's going to AA meetings now."

"Good," Lance consoled. "I hope he gets reformed and everything, because I don't want to hear about him laying a hand on you again."

"Thanks," she smiled. "Will that be all?" she asked with raised eyebrows.

"Yes, and thanks again."

He smiled and finished sipping on his coffee. A sandy haired guy stuck his head around the corner looking into Lance's office. "I heard you came up with a new phone app!" he quizzed.

"Who told you?" Lance questioned.

"You know how word gets around," the guy responded as he combed his hand through his heavily tonicked wet sand colored hair.

"Yeah, I should have known," Lance added.

He came on into the office gesturing with his left hand picking for more info. "So, is it a real biggie? Something we just GOT to have?"

"You'll have to wait and see, Simon" Lance smirked.

"Arghhh! I hate suspense," he sighed as he retreated back out of the office slightly annoyed that he could not get any details.

Lance kept doodling with his ink pen and smiling to himself thinking about Simon's attempt to pry. Then a thought hit him all of a sudden. He snatched up his cell phone. He dialed a number and listened. When the party answered, Lance began conversating.

"Hey, Daryl! I got another invention for you to register for a patent for me!"

"This you Lance? You must've got a new phone?"

"Kinda sorta. I've developed a phone with a special button on it for self-defense purposes. The phone is what I want you to register for me."

"Oh yeah!"

"Yeah."

"Sounds interesting!"

"Well, I'll give you more details a little later. I got to run to do a presentation on it in the board room right now! Holler at you later, okay?" He ended the call, placed his phone on vibrate, and dropped it in his pocket. He stepped out of his office headed to the board room.

Lance smiled as he zoomed down the highway heading home from the office in his Land Rover. His big debut of the phone app had blown the minds of the other board members just as he had imagined. When it was taken to vote, it yielded a unanimous vote for production. The six months of work he and Dexter had invested in it had really paid off. He sighed with relief now that he had a huge weight off his shoulder. He left the downtown area of Seattle and headed to the suburbs. He got the urge to have a beer so he pulled off the highway to a convenience store to grab one. He walked over to the cooler, grabbed a couple bottles of Bud Lite beer, and then walked to the snack stand and grabbed a bag of hot pork rinds. He paid for the merchandise and headed for the door. A Toyota Corolla with tinted windows sat on the driver side of his Land Rover with the engine running. A guy jumped out of the passenger side of the car rushing toward him with his right hand buried down in the crotch of his pants. Lance snatched his head around looking at the guy who responded saying, "I have a gun! Do EXACTLY as I tell you! Now, get in your vehicle and don't try anything crazy!" Lance got in his vehicle as told. The guy kept the gun pointed at him as he dashed around the front of the Land Rover and then on into the passenger side. He finally pulled his hand on out of his pants yielding the gun he had. Then he said with a grin, "Drive to the ATM down the road! You're fixin' to make a little withdrawal!" Lance put the vehicle in reverse and started down the road. The Corolla left the store trailing behind them. Within seconds, Lance's phone rang. The guy told Lance, "Answer it, but I swear to God if you give any kind of secret signal or say anything I THINK is a code, I'll kill ya!" Lance answered. It was Lisa.

"Hey, hun, where are you?"

"Uhhh…..I'm about ten minutes down the road. Why?"

"I wanted you to swing by and pick up some plates I ordered at China Kitchen," she replied.

"Okay," he answered.

A new smile engulfed his face. He had forgotten all about the self-defense app button until now. The guy frowned looking at his new smile and hollered, "What the fuck you smilin' about?"

"Oh, nothing," Lance answered and secretly depressed the "H" button on his phone as he put it down. Instantly, fog appeared and then disappeared within three seconds. Now, his old buddy, Dexter Smith, was sitting directly behind the stranger and began popping his knuckles and his neck with quick jerks to the left and right. The stranger jerked his head around in panic once he heard the popping sounds and got spooked. "What the?" he shouted as he grabbed the door handle to get out of the car. Dexter grabbed him from behind choking him savagely. Lance swerved his vehicle off the road and ran to the passenger side. He snatched the thug out of his car out of Dexter's grip and began punching him. Dexter got out and joined him beating the guy unmercifully. Lance then reached in the car, retrieved his phone, and called the cops. Afterward, he pressed the mystery "H" button again on his phone, and Dexter disappeared as soon as he punched the thug's face one last time. The guy fell over to the ground sideways. Soon, the cops came and began getting out their patrol car. Lance told them what had happened.

"What happened here, sir?" asked the officer as he walked toward Lance glancing down at the guy lying over on his side on the shoulder grass.

"Well, I was at the Quick Stop store down the road, and this guy jumped out the passenger side of a running car and ordered me to get in my car and head to an ATM where he was gonna make me get out money. However, I caught him off guard and gave him a chop to the throat with my right hand which allowed me to swerve off the road and run to his side of the car and beat the crap out of him."

"Looks like you put a whammy on him," said the officer.

He bent down to the guy who was coming back to. He pulled out some handcuffs and put them on the battered guy as he read him his rights. The other cop stood him up and loaded him in the police car. The first cop to speak then spoke once more.

"Well, my hat's off to you, sir, for a job well done," he said.

"Oh, it was nothing," Lance responded.

"I wish EVERYONE could have this same outcome with these lowlifes," he added.

"Yeah, maybe they will one day," Lance finished with a smirk on his face. In his mind, he knew he had developed the answer to worldwide safety from this sort of thing.

"So, you alright sir?" one cop questioned.

"Oh, I'm fine," Lance replied taking inventory of his clothes and knocking the loose soil off knees of his khakis.

"Well, I'm gonna write this up basically as an attempted robbery by him (pointing at the thug in the car) and a self-defense style beat down from you onto him. You can get a copy of it downtown after tomorrow afternoon, if you like."

"That'll work," responded Lance.

"Well, sir, I'm gonna let you go 'head and bounce and have yourself a good evening, okay?"

"Same to you, officer"

Lance hopped back in his Range Rover and dialed his wife.

"Hey, honey, you're not going to believe what happened!"

He gave her the run down on what happened as he zoomed on to the China Kitchen restaurant to get their dinner.

***

4

Six months passed by. The phone was now in all of the major cell phone retailers and was flying off the shelves. The Fielding-Smith App was proving to be the perfect answer to school bully situations, female defense situations, and even male defense when the time arose. The phone app began to be called the "SDF App" which was the pseudonym for the "Self-Defense" app. All one had to do is hit the "H" button on their phone in a dire situation, and the "Mr. Muscles" hero would instantly show up in a puff of smoke and would be ready to instantly get physical however the predicament required for a twenty minute time period. And the phone owner could tuck them back away by hitting the same button afterward and pay the $19.99 charge attached to their phone bill.

    Mandy Croombs, Lance's secretary, drove her Volvo sedan into her driveway. It had been a long day, and she wanted to get into the house and unwind as quick as possible. She grabbed her purse and stormed to the mailbox like usual then circled on back to her front door. "Peggy!" she shouted down the hall to see was her daughter home from school. No answer. She went to her room and seen no sign that she had come home yet. She tossed her purse on the sofa, grabbed the remote to cut on the T.V., and plopped down beside her purse. She pulled her new phone out of her purse. She just adored it even though it was a few months old. It was the new one her boss had promoted. She discovered a text message from Peggy that read: "Be home after the Science Club Meeting around 9." She had decided that pizza would be the dinner for tonight and placed an order. Her daughter could always reheat her a couple slices up in the microwave. Her husband, Jack, would be home in a half hour about the time the pizza would make it to the house. Mandy kicked off her heels and went to slip into some house clothes; her skirt was stiff and starchy and her blouse was scratchy, and she wanted out of them as quick as possible. Once she had changed into sweat pants and a T-shirt she felt totally relaxed. She plopped down on the sofa once more watching T.V. and waiting on the pizza. She heard someone at the door unlocking it. The door opened. It was Jack. "Hey, baby," he smiled. He lumbered on in and headed to their bedroom undoing his tie. "What's for dinner?" he asked.

"I figure we'd have pizza tonight. What you think?"

"PIZZA!?" Jack shouted.

"Yeah," Mandy repeated in surprise.

"I don't want no damn PIZZA!" Jack shouted.

"Well, you don't have to be so rude about it!" she scolded.

"Why can't you get in the damn kitchen and COOK me something to damn eat like a wife SHOULD?" he snapped.

Mandy exhaled in disgust.

"You've been drinking again, haven't you?"

"What the fuck does that have to do with anything?"

"I told you if you didn't straighten up once and for all we were finished! I'm NOT going to put up with you sneaking and drinking when you CAN'T seem to handle it!"

Jack came barreling to her and punched her square in the jaw on her blind side knocking her to the opposite side of the couch. He then dived on her and grabbed her by the hair. He began choking her and snarling, "Talk yo bad talk NOW! Uhhh-huh!" Mandy saw her phone just at her right arm's reach on the opposite side of Jack. She reached for it and felt for the wire mesh cage on the keypad. Once she found it, she fingered it open and depressed the "H" button. Instantly, a cloud of white smoke appeared just as it had happened in the board meeting at Galactic Corp. Then, the smoke cleared in seconds and Dexter Smith from the office in nothing but yellow Speedos, socks, and black combat boots was in front of their sofa looking furious with see-sawing chest muscles. He grabbed Jack and began punching him tirelessly while Mandy pulled away and dialed 9-1-1 on her cell. The doorbell rang. Mandy dashed to the door and snatched it open. It was the pizza deliveryman. Mandy stormed out the front door leaving the pizza guy standing there in shock at the beatdown going on her living room floor. She made it across the street to her neighbor who was attending her flowers and hysterically blabbed out what was transpiring in her house across the street. They both ran back across the street to Mandy's house. The pizza man was still standing in the driveway in the doorway getting an eyeful of the free fight he had stumbled up on. It wasn't everyday you got to witness THIS kind of drama!

 Mandy and her neighbor tore back in the house to find Jack lying in the living room floor knocked out with Dexter kneeling over him all sweaty and breathing heavy. The cops drove up with screeching tires as they pulled onto the curbside and dashed to the door. They yielded their guns and demanded, "Freeze, sir! Put your hands up!" Dexter obeyed and put his hands up as he sat on the floor. They both briskly walked his direction as Mandy attempted to tell them her story, yet Dexter vanished into the thin air. Jack was still unresponsive on the floor. Mandy began trying to explain.

"What the hell?!" one shouted as they both jerked their heads from left to right looking for Dexter. They both rubbed their eyes and aimed their pistols as they made arc swings around in the living room still searching for the muscled up black dude that was just in the room.

"Officers, I can explain!" she began.

"Where did he go?" the other cop yelled.

"Well, you're not gonna believe me, but he went back into the phone." Mandy answered.

"Say what?" both cops chopped with puzzled looks.

"Yeah, my boss and the black dude you just seen invented the new self-defense phone app that came out a few months ago; and my boss, the black guy's partner, put the black guy's image into the phone as a superhero-like man who comes when you need him to fight for you. Well, my husband attacked me after going on a drinking binge with his friends tonight, and I had to use it."

"So you're sayin' that your boss and the black dude that just vanished invented a new phone app?" one cop asked with a raised eyebrow.

"Yes, my boss is in the R & D department at Galactic Electronics." she replied.

"But this is your husband lying here?"

"Yes."

"That dude really beat the crap out of him." one cop commented.

"Fill me in on how this thing works then and how this BLACK GUY winds up disappearing before our very eyes. " the cop demanded.

"Mam, your pizza?" the pizza guy interrupted.

Mandy briskly walked to a bookshelf and got a twenty she had lying there for the pizza. She handed it to him and told him to keep the change. He was very interested to see and hear all the details, and knew that he needed to be getting back to work, but today was an exception. You didn't see this type of thing every day. He figured that he had a valid excuse to be away for a longer time, and so sat on a bar stool listening about the app details as Mandy explained how everything unfolded to the cops.

"Well, I got home about an hour ago like normal, changed clothes, and ordered a pizza for our dinner. My husband came home and had been drinking AGAIN, and told me quote unquote 'I don't want no goddamn pizza! Why don't you get in the kitchen and cook like a wife's supposed to!' And then he walked over to me and stole me while I wasn't watchin' and punched me in my jaw as I was fussing about him drinking again. After that he dived on me and started choking me telling me "Talk yo bad talk NOW!" It was then that I reached for the closest thing I could grab which was my new phone, and I mashed this button here. (She showed both cops the button.) The black dude instantly showed up in a big cloud of smoke once I mashed the button and then grabbed Jack and started beating the crap out of him."

"Well, that's a VERY interesting story!" one cop admitted. He then added, "We're going to write up an incident report and you can get a copy tomorrow sometime. Do you think you're going to be alright?"

"Yes, I'll stay with my mom and dad tonight, thank you."

"I'm sorry you had to go through this, Miss," a cop advised.

"Yeah, sometimes you just can't get SOME men to do the right thing, I guess."

One cop grabbed Jack and put handcuffs on him and began reading him his rights. Jack looked like a monster with his black eye and bloody mouth and began regaining consciousness from being stood to his feet. The cop finished reading him his rights and hauled him on out the door. The other cop was asking general questions and quickly writing up the report on the clipboard.   The pizza guy finally took off and headed back.

"Well, mam, we're going to say Good Night. Take it easy." The cops remarked.

"Oh, you guys can have this pizza if you want on me. My appetite is gone now." Mandy said as she handed one of the cops the whole pizza and closed the door. Blowing air from her bottom lip making her brunette bang flutter, she walked over to her husband's bar and fixed her a shot of bourbon and Coke. "He's gon turn ME into a damn drinker!" he snapped as she sipped some of the drink. Then she came back over to the sofa and plopped down beside her neighbor. For the next two hours they chatted about the occurrence and before they knew it, her daughter was walking in the door. Oddly, she had a taste for pizza like her mom had had, so Mandy peeled off money to her so she could order another one. The neighbor let the daughter's arrival be a good time to return home and offered her dinner once more and asked if there was anything else she could do before she left. Mandy politely refused all offers, seen her out the door, and then phoned Lance and told him what had occurred. She let him know that she was grateful for the new phone app and felt he should hear her testimonial while it was fresh on her mind.

***

5

    One whole year rolled by. Millions and millions of the phones were in the hands of consumers. Hundreds of thousands of testimonials were flooding social media channels discussing how they were assisted by some strange bald black man in Speedos with bulging muscles. People were taking selfies with them and the pictures were bombarding the social media platforms. Plus, there were women who were mashing the button just to see and admire their Mr. Muscles' photogenic body and try to make out with him. But these women would have to keep mashing the button over and over thus piling up a bunch of charges on their bill. Consequently, cell phone bills were skyrocketing from all of the curious patrons, all of the needy patrons, and all of the freaky patrons all across the country.

Mrs. Owens arrived at her house with her new cell phone. She could hardly wait to see her black hero again who had appeared in her classroom. In her mind of what a man should look like, he was "perfect." She unlocked her door, stepped into the house, placed her attache case down beside the door, grabbed her old phone out of her purse, and then flung the purse on the sofa. Then she googled the hit "Freaky With You" by Silk on her old cell phone. Once she found it, she got it set to play within a moment's notice. She poured up two glasses of wine. Then she went, took a quick shower, and changed into a red negligee and red pumps. She put on makeup, red lipstick, and sprayed on a little White Diamonds too. She had to have everything just right. She went back to the dining room and retrieved the new phone out of her purse and placed it beside the old one on the coffee table in front of her once she seated herself in the living room. She took a sip of wine and sighed. It was time. She pressed the "H" button on the new phone. Mr. Muscles appeared just as anticipated with a cloud of white smoke that cleared within seconds. She then pressed "play" on her old phone, and "Freaky With You" started playing. She grabbed Mr. Muscles by the hand and led him to the love seat. Then, she pushed him down and then sat down in his lap. She could feel his stiff manliness under her. She teased, "Is that a banana in your trunks, or are you just excited to see me?"

"It's a banana alright." He replied sarcastically licking his lips with a smile.

They began kissing passionately. The music was playing and the mood was right. Mrs. Owens was now in hog heaven. Her main reason for buying the new phone had come to fruition. He threw her down and began undressing her. Then he disappeared without warning. Mrs. Owens desperately sprang up to grab the new phone and mash the button once more to bring her lover back. She was determined not to let one single minute slip away. She grabbed it and accidentally dropped it right inside one of the glasses of wine that sat on the coffee table. "Shitt!!!!" she hollered as she scampered to retrieve it in utter disbelief of the luck she was having. Her effort to save it was to no avail since the phone was completely dead. The love affair that she had started had come to an abrupt halt. "DAMN….DAMN…..DAMNN!!!" she hollered plopping back down on the loveseat in disgust. She grabbed the wine glass and flung it across the living room striking the opposite wall leaving a lavender colored streak on the wall from the wine that splattered. After a few moments, she sighed, sucked her teeth, and said, "Oh well….back to the drawing board."

***

6

    Lance sat watching the evening news. Lisa had gone to pick up some Chinese food take out from the China Kitchen restaurant. Various things were on the news: Robberies, wrecks, murders, natural disasters, you name it. Lance heard them and just grimaced feeling a sense of helplessness. Then, he heard about a three month old baby that had gotten left in the back seat of a hot car during the current hot summer temperatures. He became frustrated trying to figure out how a parent could be so slack. This had turned into a very terrible pattern nationwide. An idea struck Lance, and he smiled. He now had work to do in his lab. And his old buddy, Dexter Smith, would come in handy.

The next morning, Lance strolled into the front door of Galactic Electronics whistling with his briefcase. He had worked on his new idea until late the night before and was excited to show it off in the morning board meeting. He marched into his office, sat his briefcase down, and dialed up his friend, Daryl.

"Yeah, Daryl?"

"Hey, Lance, my man!"

"I got another one for ya!"

"Oh yeah?"

"Yeah, and I want you to register a patent on it."

"Give me a clue, what's it good for?"

"It's a solution to all of this leaving kids in hot cars that's goin' on!"

"Oh REALLY?"

"Yeah."

"I'll have Mandy fax the patent application to you later, okay?"

"Alright. I'll be here."

Lance got up and headed to the coffee pot. He fixed a cup and went back to his desk meditating on his new invention until time for the board meeting. He had payed a woman with a small one year old baby $75.00 the evening before to help him conduct a trial an error of a new kid detection system in vehicles. She was to start "Zip Chatting" him by phone at 9:15 AM and broadcasting what occurred as she stepped away from her vehicle leaving her child in the car seat. The members of the board would observe the video from the office on the computer screen.

The board meeting began promptly and on schedule at 9 A.M. Lance stood up and turned on the computer monitor. Then he began speaking.

"Good Morning, Members of the Board. I have a little show and tell for you this morning before you get into your normal numbers talking. Betty, are you out there?"

"Yes, Mr. Fielding." A young white woman was now sitting in her SUV looking into her phone's camera.

"Okay, Betty. Now, you can get out and begin walking away from your SUV."

"Okay." She got out and started walking away from the vehicle. All of a sudden a cloud of smoke appeared. The board members started asking "What's going on?" Before they could all react they saw Dexter Williams appear before their eyes on the screen. He began to chastise the woman.

"Hey, GET back over here and get that baby out of this car!" he hollered in her direction. She stopped in utter surprise wondering where the man had come from. She obeyed and went to retrieve her kid. Dexter told her the ramifications of her actions.

"Now, you'll have an extra $35.00 charge when you get your next decal, and if I have to tell you AGAIN, it'll cost you an extra $55.00! Got me?" He then turned and walked away vanishing in the morning sun rays of the parking lot.

Everyone at the office began applauding. Just like he expected, Lance then knew he could get the funding with no problem. Questions began to erupt from the board members.

"So, Lance, Ole Dexter's likeness is going to appear whenever any child is left in the car seat?"

"Yes."

"Well, how is it that he knows?"

"A weight sensor in the child seat detects that the kid is still in the seat and triggers triggers Dexter's hieroglyphic likeness to appear and thus save the baby by reprimanding the parent." Lance clarified.

"I must say…This is a very novel creation!" a board member stated.

"Astounding!" jeered yet another board member.

***

7

One year passed.   It was blistering hot. People were walking frantically out of the mall to their cars and the other way around to escape the heat. A Chevy Tahoe rolled into a parking spot. The driver jumped out with her phone in her hand busy tinkering with it. She walked toward the mall with her blonde hair tied behind her head in a ponytail. Suddenly, she felt a sharp tug on her ear lobe. She hollered and froze in pain and anguish as she heard a voice say, "If you don't get yo ass back over there and get that kid outta that car!" She spun around to see a strange bald black guy in Speedos who was easy on the eyes. He held on to her ear as she frowned in pain walking back to the Tahoe. She hit the keyless entry and opened the back door. Then, she reached in and got her three year old son out of the baby seat. She told the black guy, "I got to get his stroller out the back. Hold him." She handed the boy to the black guy and headed to the rear of the Tahoe. The guy took him in his arms and waited until she got the stroller out. Once she had the boy placed in it, he said in a pissed off tone, "Don't let me have to tell yo ass to get this boy out the car AGAIN! There'll be a $35 fee added to your registration fee when you get your next decal! If you do it again, there'll be ANOTHER $55 fee! And so on. And so on. Got me!?" Wearing a frown, the woman answered, "Yes! "

The black guy turned away and began walking away from her and said, "Now, have a good day!" He vanished as he was stepping leaving the woman gazing around to see where he went and if anyone else seen him. She hurried inside the mall pushing her son in the stroller.

Over the course of the next few months, similar accounts happened to drivers who would forget to take their young kids out of the child seat. Dexter was continuing to catch people off guard and save their kid's life before they got away from the vehicle. Word was getting around about the strange black man like wildfire. Some were overly curious to see what he looked like but didn't want to pay the $35 fee when they renewed their decal just to see him.

***

8

    Mrs. Owens rolled over in bed and became semi-conscious to the sound of down pouring rain outside her room window. She smiled and gripped her pillow tightly keeping her eyes closed. Her foot struck a masculine leg under the sheet. "Don't you just love the sound of rain, babe?" she whispered. "Yeahhhhh, "the male voice answered. Then, her eyes opened, and she smiled into the eyes of Mr. Muscles. He reached over and grabbed a handful of her buttocks and then smacked them in the darkness of the room. The sting turned her on. "How 'bout we do a lil sump'm-sump'm?" he asked. "You know I'm down with that." she cooed. He rolled over on top of her, and they tied up kissing passionately while the rain poured down heavy outside. Mr. Muscles being already naked like she was inserted his stiff manhood into her, and she could hardly wait. She caressed his back as he began pushing inside of her and pulling back at a steady rhythm. After a few moments, the lightning struck, and she jumped. She was now fully alert to her surroundings and to her surprise all ALONE in bed. She had been dreaming about being in bed with Mr. Muscles. "Well, I'll be DAMNED!" she breathed. She fell backwards onto her pillow in disgust, and for a few moments she kicked hysterically as if she were having a temper tantrum. She grabbed her other pillow beside her and screamed into it as she continued kicking. "I have GOT to get another one of those damn phones!" she shrieked finally.

***

9

Destiny Meadows pulled into the driveway of an upscale home in Tumwater. She strutted on inside the home with her blonde hair bouncing and her heels clogging a steady tune. She was a few minutes ahead of the couple that wanted to see the house. She walked inside and turned the air conditioning on to get it cooling off some for her clients. She heard a door close outside and peeped out the window in the living room. It was a guy in a Polo shirt and Khakis but his wife wasn't with him like he had said she would be. She composed herself tugging at her skirt and waited for him to ring the doorbell. He rang it. She opened it and greeted him.

"Helloooooo, Mr. Watkins! Did your wife not come?"

"Oh, she's coming. She's coming from her office in Everett."

"She's about five minutes away, though."

"Okay, great! How was your day?"

"Oh, fine. I'm ready to unwind. It's been a long day."

"I bet! What kind of work do you do?"

"Oh, I'm a car sales manager."

"Where at?"

"At the Ford dealership in Tacoma."

"Oh!"

"Well, you're right around the corner from here."

"Yep."

Destiny stopped talking and stood quiet for a few moments gathering her thoughts. She didn't want to start showing the house without the wife present. It wouldn't be tactful. That was one of the Dos and Don'ts of being a realtor. Plus, she was getting a little apprehensive with it just being her and this strange middle aged man in the house. She sighed and began talking again while gazing toward the front window.

"How do you like what you see so far?"

"Oh, I LOOOOVVVVVEEEEEE what I see thus far!" the guy said with a grin and added, "and I like the house too."

Destiny paused unsure of what to say next.

Then she said, "Mr. Watkins, you're a handful aren't ya?"

He smiled and said, "What's takin' that wife of mine so long, I wonder?"

Destiny was glad he asked about her before she had to. Then she told him, "Why not call her and see where exactly she's at?"

"Good idea, " he answered winking at her as he snatched his phone out of his pocket proceeding to punch in the numbers.

Within seconds he was talking.

"Hey, girl…..Where you at?........Five minutes away!....... Okayyyyy, you got us tied up here waitin' on you! But take ya time! We're tryin' not to start the party without ya! (winking again at Melody again)…….. Bye." He poked the phone ending the call.

Melody grabbed her phone and told the guy she had to phone the office for a moment while they waited for his wife.

She phoned and walked to the kitchen area talking. Mr. Watkins walked over and locked the doorknob and the bolt lock. Then he walked to another room and hid himself. He heard Melody laugh and say, "Okay. I'll call you back in a few." Her voice echoed quite well since the house was empty. Then she called out, "Mr. Watkins?........Mr. Watkins? Where'd you go?" He could tell she was heading his direction and he stood just inside one of the bedrooms waiting for her behind the door. When she walked in he grabbed her. She screamed, and he warned her that no one could hear them. He forced her down onto the carpet with a choke hold on her. She kicked him in the shin, and he backhanded her. She tumbled to the floor. She felt her sore mouth and got blood on her finger from her busted lip. She wiped her hand on the carpet. She thought about her new phone. She had dropped in her coat pocket when she called the office instead of walking back to her purse. She hurried her hand into her skirt suit pocket feeling for her phone and slipped her finger under the cage where the "H" button was located. The guy snatched her bra loose exposing her breasts as soon as she had depressed the button. Smoke appeared in the room restricting all visibility. Then it cleared just as fast as it came, and Destiny and Mr. Watkins lay on the carpet in front of a muscular bald black dude who stood angrily looking down at Mr. Watkins with his chest muscles see-sawing. He dived on him punching him mercilessly as Melody scampered away and ran to another room screaming and dialing on her cell phone. The muscular dude kept wailing on him beating the crap out of him for a good fifteen minutes.

Then he held out his hand, snapped his fingers, and some handcuffs appeared out of nowhere into his hands. He placed them on Mr. Watkins and tugged him near the front doorway. "Sit yo ass down!" he demanded as he shoved him to the floor. Then he addressed Melody.

"The cops are on the way, mam. I used my telekinesis to dial them. Sorry you had to endure this! This piece of shit won't be trying that again, I betcha!" He held out his hand toward her as she frantically tried to pull her jacket together to hide her bare breasts. The buttons of her skirt suit jacket had been snatched off when Mr. Watkins snatched it open and tore her bra. Destiny stretched her eyes, and her jaw dropped in disbelief of seeing this guy reaching at her bosom even though it was only for a second. Then, magically, a beach towel appeared in his extended hand. "Here, cover up with this!" he told her. She was relieved once again and took the towel and wrapped up in it.

Destiny was in awe of how the new app feature helped her in her dilemma. Boy did she have a story to tell her girlfriends at the office when she got there. And she couldn't wait to describe how good her rescuer looked also.

The cops soon came to a stop outside the house and jumped out running to the door. They knocked and twisted it open.

Destiny had already unlocked it. Mr. Muscles told what happened.

"Officer, this lady was accosted by this guy here who tried to rape her, and she mashed the app button to call for my help. I came and whooped his ass, and here he sits."

"Oh, so your phone has that new app I'm hearing so much about?" one of the cops asked Destiny.

"Yeah," Destiny replied tugging on the beach towel around her neck.

He nodded his head with his bottom lip protruded giving it his nod of approval.

"Well, let's load his ass up!" the other cop said as he reached for Mr. Watson's cuffed hands. He stood him up and led him outside reading him his rights..... "You have the right to remain silent. Anything you say may be held against you in a court of law..... The cop's voice faded once he got the guy outside the door.

The cop left inside began talking.

"Mam, I'm going to briefly jot down what happened here, but you can come downtown and get a copy of the report later on."

Mr. Muscles vanished instantaneously into the thin air. "What the?" the cop hollered looking from side to side. Both he and Destiny snatched their heads from side to side trying to see where he went. "I guess he left just like he came, huh?" he asked. They had never witnessed anything magical like this before. "What's your full name mam?"

"Destiny Meadows."

"And this guy ripped your clothes, …. struck you in the faced……but he didn't actually rape you did he?"

"No, he was ABOUT to until I thought about my phone app and mashed the secret button which summons the black guy you seen."

"I see. Well, you okay, mam?"

"Yeah, I'm okay."

"So we don't need to do a rape kit on you at all?"

"No, you don't." Destiny reassured him with an expression of finality about the notion.

"Okayyyy. Well, we're done here. I'll leave it with you then."

"Oh, I'm leaving when you leave. I was showing this HOUSE to that creep. I'm a realtor."

"Ohhhh, okay." The cop said putting the last piece of the puzzle together.

They got into their respective vehicles and crank up. The cops backed out freeing the driveway so Destiny could do the same. As she backed out, it finally occurred to her that the guy's wife never showed. It was all an act! It gave her chills to realize how dangerous her profession really was.

***

10

All over the country babies were both being saved, and criminals were being dealt with more quickly because of the Fielding-Smith App. It was really proving to be a novel and timely invention. The thirty-five dollar fee was nothing to pay extra when renewing one's tag decal for the year compared with the other foreboding alternative of losing a child due to negligence. Cops were digging the app too since they were getting some help on the streets by the infamous "black muscle man" who'd have their culprits already handcuffed by the time they showed up on the scene. Lots of single women all over the country and English speaking world were activating the button just to see the mysterious guy show up just so they could gaze at him and try to seduce him. The fact that he only showed for twenty minutes if there was no real need for him did not matter; they only mashed the button again and again to keep him coming. A person's budget was the only constraint in making the guy show up countless times. Reports began to hit the news that women were having sex with the "Mr. Muscles" guy, and that some were even getting pregnant. Photos that people took of him all revealed the same guy, Dexter Smith.

***

11

It had been two weeks since Mrs. Owens accidentally dropped her phone in the glass of wine ruining her plans with Mr. Muscles. She switched in her signature style of walking into the local Cell Phones R Us store and stepped up to the desk.

"Yes, may I help you?" asked the young black female clerk.

"Yes, I want the new phone with the Self Defense App."

"Okayyyyyy, I'll be glad to get you one." she stated reaching down to unlock the glass case where the phones were displayed.

"I had one, but I dropped it in water; now I'm forced to buy another." Mrs. Owens remarked.

"Oh, I know you must have cringed when THAT happened!" the girl snapped.

"Yeah, girl! I had to come get ANOTHER one of THEM phones and QUICK! That man that shows up is sump'm serious! You seen him yet?" Mrs. Owens asked.

"No, but I've heard about him and seen photos of him!" the girl commented.

"Girlllllll, you betta sneak back there and mash that button YOSELF on one of these phones!" Mrs. Owens insisted as she reached into her purse and pulled out her debit card.

The girl put a box on the glass countertop.

"Now, mam, let me show you the features of this phone."

"Oh, they showed me the features just last week when I bought the other one I told you about. So, I already know." Mrs. Owens clarified.

She handed the girl her debit card and paid for the phone, and the girl handed her card back to her. The sale was complete.

"Bye, mam, I hope you have better luck with THIS one." the girl mentioned.

"Thank you. Oh, I'll be EXTRA careful THIS time." Mrs. Owens smiled looking back.

She made her exit of the store out into the parking lot. It was blistering hot. As she reached for her shades in her purse on the sidewalk outside, she heard a familiar voice calling out to her.

"Heyyyyyy, Miss Owens!"

She looked toward the sound. It was a young black boy from her class walking beside an older guy presumedly his dad.

"Hey, who is that?" she hollered squinting in the sunlight.

"Antwan!" he replied.

"Oh, hey Antwan! What are you up to today?"

"My dad and I are going shoe shopping." he answered as they both closed in on her.

The both came face to face with her. His dad was a real looker. He had a handsome face and build with pretty white teeth. He made her want to fight the beaming sun to chat for a much longer duration. It had been awhile since she had a relationship, and she was trying to remain ladylike and fight her animal desires and not appear like a ravenous wolf.

"Hi, I'm Luther Pendergrass, Antwan's dad." the man said holding out his hand. She had never met him. Antwan's mom had registered him to school by herself and had come to the school without him the few times she had shown.

"And I'm Peggy Owens. Nice to meet you." she responded with her hand extended to meet his.

"Where's his mom, Miss Margaret? She's the one I'm used to seeing all the time at the school." Mrs. Owens asked.

Luther sighed and then said, "We separated."

"Oh, Noooooo!" Mrs. Owens exclaimed with a sad look on her face while she jumped for joy on the inside.

"Yeah, it's one of those things."

"I'm so sorry to hear thaaat! Well, if there's anything I can do to help, let me know. Okay?" she said in a way as regretful as she could muster. In her mind she was saying, "If you need me to come over and help you forget about her just holla."

"Thanks for caring. I guess me and my boy will mozie on then."

"Mozie on?" Antwan questioned as he looked up at him squinting from the sun's rays.

"Yeah, MOZIE on! You never heard of that?" Antwan's dad quizzed as they walked on toward the mall entrance. Mrs. Owens heard Antwan answer, "No" as she stepped on toward her car. Once she got to it, she noticed that it looked a little one-sided. She walked to the passenger side and looked. There it was……….a flat tire! Immediately, she thought of her new phone. She grabbed it and mashed the H button under the cage. Instantly, there was white smoke which cleared leaving Mr. Muscles gazing at her with squinted eyes in the bright sunlight.

"Let me guess……You need your tire changed?"

"You guessed it."

"Aw hell, NAW! Now, I ain't down with no damn tire changing! This is a SELF DEFENSE APP!……not a car repair app! You need to call one of those auto clubs……AAA or something!"

"So you ain't gon fix my tire?"

"Helllll Naw I ain't gon fix yo tire! Sheeiiitttt. The only things I do that start with an 'f' are fuss, fight, frighten, and fuck. (He grabbed a handful of his red Speedo on the last word while thrusting his crotch forward.) And fixin' ain't one of 'em!" he finished shaking his head as he brought it close to her.

"Well, fuck YOU then, you bastard! You can get yo black ass back in this damn phone, then!"

"Bitch, you ain't said nuttin but a word!"

He went reaching for the phone himself to mash the button over Mrs. Owens shoulder. She wrestled her way around in his direction as she sat in her driver seat and slapped him. He grabbed his face with a frown, and then they began staring at one other with angry faces. Then they locked up like two pit bulls, but instead of fighting they were ferociously kissing. They fell across the seat and center console lost in their own passion and desire. For a few moments they failed neither to realize how hot it was inside the car nor how uncomfortable it was for the center console to be pressing them in their backs. Mrs. Owens reached over and crank up the car without losing a stride in necking with her Mr. Muscles. The cold air began to cool off the interior while they continued making out. Being proactive this time around, she thought about the time that had passed and reached into her pocket to make sure she had the new phone in her reach. Just as she estimated, he vanished. It had been twenty minutes. She quickly depressed the button once more, and he reappeared after a second of smoke coming and clearing away like the times before. She was determined to see this thing through THIS TIME even if she had to mash the button ten times straight! She told him that she was going to head home and that they'd finish their business at her place where it was more comfortable. She began frantically driving as best she could with her hazard lights flashing on to her house with the flat tire. The car was leaning to the right and pulling to the right, but she managed to keep it on the road driving the best she could manage. In five minutes they were there. They both leaped out and dashed inside the house after Peggy frantically unlocked the door. They began feverishly stripping inside the doorway. It was extremely easy for Mr. Muscles since he only had three pieces to take off: his Speedo, his socks, and his shoes. They both dashed to Mrs. Owens' bedroom with her in the lead holding her new cell phone. He vanished again, but she quickly mashed the button since she knew what was going to happen. She responded so quick this time, that there was no white smoke. Once back, he snatched off the Speedo, socks, and shoes again and then, it was "Lights, Camera, Action", so to speak. The two fell into a wild animal-like romp of nonstop sex while surrendering to their base carnal desires. It had been over six months since she been in a relationship, and she had a lot of pent up sexual tension bound up inside. Rhythmic body clapping sounds echoed from the bedroom like a boomerang to the other parts of the house as they indulged themselves in canine style sexual activity. Moans, shouts, and expletives flew from their mouths uncontrollably. Mrs. Owens couldn't believe she was finally sexing the Genie man from the cell phone that she had been lusting for ever since she lay eyes on him for the very first time inside her classroom and found herself climaxing over and over until she was as limp as a noodle.

  Finally, after thirty minutes of invigorating sex, they both collapsed side by side in bed. The sheets were all tangled and their bodies were all covered in sweat which began cooling them off in a refreshing way as they lay there under the ceiling fan. Mrs. Owens gently caressed Mr. Muscles' well sculpted moist abs with her hand. Within seconds, her lover disappeared once again, but Mrs. Owens just smiled this time. She had accomplished her goal at last and had no need to retrieve him as she had done before. But his departure did trigger her to think of the one hundred bucks she had piled up in charges between Peter Fielding's phone in her classroom to the charges of her own that she had incurred while retrieving "Mr. Muscles" for this sexual showdown. *"Oh well, it was worth every dime!"* she thought to herself. She closed her eyes and formed a smile.

***

12

It was like any other day in Seattle. Despite the drizzling rain, it was business as usual at Bank of America. Clients came and went busy doing their morning routines. Three guys walked in conversing with one another in low voices about the weather. It always had a special way of igniting conversation between strangers. The guys split up going into different lines to different bank tellers. Moving at a snail pace, the separate lines brought the three fellows closer and closer to their respective windows. Once the first one reached his window, he threatened the teller.

"Do exactly as I say, and you won't die, okay?"

"Okay," she answered.

The other two guys snatched pistols out of their coat pockets and jumped out of line brandishing them while prancing around and scoffing at the folks in the bank vicinity telling them to get seated on the floor. One of the armed guys backhanded a guy that was moving to slow in seating himself on the floor.

All of a sudden there was lots of white smoke everywhere. Then, it cleared without a trace. Over a dozen identical muscular black guys in red Speedos were now standing amongst the crowd looking around angrily. Several of them doubled up and pounced upon the three culprits and began what looked like a bar brawl. Lots of punching ensued with the culprits constantly falling down while hollering in pain and agony. After ten minutes of roughing them up, the muscular dozen huddled the three guys up in a seated position on the floor. Three of the one dozen muscular heroes magically produced sets of handcuffs in their hands which they used to handcuff the robbers.

The cops showed up after a few minutes to find the worst of the work already done. They grabbed the robbers up and began reading them their rights. Then, the led them to their patrol cars. The crew of bald muscle men were heralded and applauded by the happy bank clients. Some wanted their autographs and desired to take selfies with them. Within a couple minutes, the dozen heroes vanished just as quickly as they had appeared. The bank clients responsible for secretly summoning help on their phones began revealing themselves to the others. They shared their stories later on the Evening News on how they saved everyone with the infamous Fielding-Smith App.

***

13

From coast to coast across America, folks were talking about the black muscle man from the phone. He had reached super hero status. Lots of criminals were being rounded up and lots of babies were being saved all across the nation. Fees were racking up by people who consciously needed help. The county governments were collecting a sizable sum of revenue that could be used to fund needy projects. Along with the consciously created fees, some fees were piling up from kids playing with the phone and women seeking the muscle man for ulterior motives.

Dexter's wife, Lashonda, sat on her sofa at her house back home in Georgia watching the evening news. After Dexter's death, she found it more relaxing to return to her hometown of Statesboro to be next door to her mom who said she would help her with the three kids. She was getting a tidy sum in royalty payments from Dexter's share of Fielding-Smith app; her and her kids were getting well taken care of. All of a sudden she saw a news flash about the infamous "Mr. Muscles" that was coming to people's rescue with the pressing of a button. She stood up staring at the T.V. She couldn't believe who she was seeing. It was her husband, Dexter! But how! Her jaw dropped and she dropped back down on the sofa. She listened to all of the hoopla about how he was helping hundreds of thousands out of danger and saving lots of kids in car seats on a daily basis. What was disturbing was the list of claims of women having sex with her husband and others that were getting pregnant by him. She pulled out her cell phone and began dialing.

"Hello."

"Yes, Lance, this is Lashonda. Why didn't you ask for my approval for using Dexter's image in this app thing?"

"Well, we figured you wouldn't have a problem with your husband being a hero. Who wouldn't want to be a national hero?"

"He's screwing women too! Tell me how THAT part relates to being a hero!"

"Well, a person's real personality will shine through. He was a very virile dude until he met you, you know. We have no control over his strong testosterone. When he met you, it tamed him down and made him give up being a player."

"I should sue for you guys using his likeness like this without asking me first, Lance. That just doesn't sit well with me!"

"Okay. You're right! We should have asked you! We just assumed you'd be overjoyed to have your husband immortalized as a hero is all. How 'bout I approach the board with a proposal to paying some extra royalties to you? To the tune of another four grand a month?"

"Well, that sounds incredible, but it's just the principle of the thing, you know!"

"Alright...I'm sorry! But , I'll approach the board tomorrow about more compensation for you for the use of Dexter's likeness. I don't think I'll have a problem getting it approved. It's the least we can do for your inconvenience."

"Well, I appreciate it. But do you realize you all could have been sued BIG TIME? I'm only being nice about it 'cause I know you'd catch a lot of heat for it, and we go too far back, Lance."

"That we do, Lashonda! And I apologize for our intactfulness."

They both hung up.

Lashonda pulled her own phone out that she had purchased just the prior week. She had been hearing about how some hieroglyphic dude would come to your rescue, but never did she realize that it was her late husband. She gave the 'H' button under the little cage a push. Whoof! White smoke was everywhere in a split second, but it cleared just as fast as it came. Now, she was staring at her late husband. She couldn't believe her eyes. "Dexter?" she whispered.

"Lashonda?" he responded.

"You mean, you remember me?" she asked.

"Yes, I remember you. I couldn't point blank request to see you in this particular realm I'm in. I could only see people as that button would summon me to their presence."

"Wow! This is SO bizarre!"

"You can say THAT again!"

She walked up to him and slowly touched him. The expression on her face was utter disbelief. "I can FEEL you", she gasped.

"And I can feel you, too," Dexter uttered.

"So, what's the deal with all those claims of women mashing the button just to have sex with you and these pregnancy claims popping up everywhere that I'm seeing on the news? Any truth to any of that?"

"Baby, I didn't know if I'd ever see you again. I promise those women mean nothin' to me."

"So those claims are true?"

He sighed and dropped his head responding, "I'm afraid they are."

"Men! Even from beyond the grave they break your heart!" she shrieked.

"I'm sorry, BABY! You know you're the only woman I truly love! Now that I've found you again, I promise I'll change my ways!"

"Well, I can't expect to hold you to THAT promise considering!"

"Considering what......that I'm not real?"

"Yeah. You're constantly gonna be getting called by women in danger and women who negligently leave their kids in cars. You can't withstand all of THAT temptation, so don't even lie to me and say you will."

"I promised you when I married you and kept that promise the whole five years of our marriage, so I damn well can do it now!" he pleaded.

"I wish things were like they were before so BAD! I miss you terribly, Dexter."

"And I miss you too, " he said hugging her.

They stood hugging one another with tears falling until he vanished. Lashonda looked around in amazement. She pulled her phone out and mashed the button once more. The smoke came and went again leaving Dexter in front of her again.

"What happened?" she asked.

"The app times you out at twenty minutes if there is no physical altercation for me to solve. If there IS a physical urgency, I vanish twenty minutes after the culprit is subdued."

"In other words, I have……..(she looked at her watch)….about fifteen minutes before you disappear again?"

"You got it."

"I want to make love to you again."

"Hey, we can do it! You just have to keep the phone near and keep mashing the button every twenty minutes!"

"That'll work?"

"It worked last week!" he quipped before the thought about it.

"So, you were fucking some woman just last week, huh?"

Dexter was short for words. He dropped his head again. Then he lifted it clasping one fist inside the other palm approaching her.

"Baby, yeah I fucked a woman, but I was thinking about YOU! Yeah, I've FUCKED other women, but I make LOVE to you. There IS a difference! You FUCK someone you could care less about, but you make LOVE to someone you really cherish and value. "

Lashonda was just looking at him quietly.

"It's like this for example……Baby, you're like a Mercedes to me. I pet YOU……keep you washed and waxed…….drive you very delicately and drive you on fancy occasions. These other skeezers are nothin' more than American lemons…..has been cars…..get by cars."

He disappeared again. LaShonda brought him back. Then she started lightly clapping her hands.

"What?" he quizzed her.

"That's for that speech you just gave. It was Oscar worthy." She insisted.

"Well, it was the TRUTH!"

"I'll be in the doctor's office the NEXT time I summon you so you can be tested before WE do ANYTHING! You ain't giving ME no V.D.!" Lashonda said with a sarcastic attitude.

"How are the kids doing?" he asked.

"Junior and Nicki are fine. I'll walk over to mom's house and let you see them. They're taking their afternoon nap right now." she said looking at her watch.

"So, it's been a whole year or more now. You ain't got another man yet?" he asked.

"Maybe……Maybe not. "

"Oh, so what does THAT mean?"

"Let's just say, I've FUCKED about thirteen guys, but I haven't made love to ANY of them. See, I make love to you, but I FUCKED them! I guess we're even Steven then."

"You got jokes, I see."

"No, I got what you call medicine, and I just gave you a dose of your own medicine. How's it taste?"

Dexter just stood there looking at her while she gazed at him. Then he disappeared again. She decided to wait to bring him back. She needed time to cool off.

***

14

A Dodge Caravan drove into the parking lot of a local high school. The driver got out with his briefcase in his hand. He ran his hand through his brown hair and closed the door. He made a couple steps toward the school and got a terrible surprise. Standing in the sunrays of the morning sunrise and immediately in front of this guy was a shiny headed black guy with well-cut muscles. He caught this guy right in the stomach and made him buckle. The guy hollered "What the FUCK?" as he struggled to stand back up. The muscular dude hollered at him.

"Get yo ass back over there and get that baby out of that damn car!" he demanded.

The guy slammed his palm to his mouth in shock. "Oh my God!" he hollered. He dashed back to the van clicking it unlock with his clicker. He snatched open the rear door and his baby was smiling at him. He whirled back around to the black dude.

"I swear to God....I thought I had dropped him off to daycare!" he gasped.

"Well, NOW, yo ass is gonna have a $25 dollar fee added to your registration come tag time, and if it happens AGAIN, it'll be $35.00. So, don't let the shit happen again! You got me?" The black dude reprimanded.

The muscular dude went stepping away with his head shining in the morning sunlight. Now, the teacher had to call his office. He snatched his phone out of his pocket.

"Yeah, Melanie, uh…This is Charles Sinclair. I'm running a little late this morning. I'll need someone to watch my class for about a half hour to forty-five minutes this morning."

"Okay, Mr. Sinclair." She responded. He looked in the direction of the bald black guy but seen no one. He thought for a moment about how tragic this could have been if it hadn't been for the black stranger. Yeah, he had gotten punched in the gut, but that was nothing compared to the other alternative of losing his son. So this is the guy everybody was talking about. He hadn't seen it for himself until now. He didn't regret the $25 fee he had to pay one bit. Losing his son was something he didn't want to even think about.

***

15

Mrs. Owens stepped outside to gaze at her car. She still had to get the flat tire repaired that Mr. Muscles had declined to fix. If he wasn't so good in bed, she would be pissed with him right now. She walked over to her car and kicked the flat tire. She reached into her pocket to pull out her cell and heard a voice holler.

"Hey, Miss Owens!"

She looked around to see Antwan reaching out his window as his dad slowed the car up by her driveway.

"Hey! Yall must live 'round here somewhere?" she asked.

"Yeah, on the next street." Antwan answered.

"Oh, okay."

"Need help with your tire?" his dad asked.

"Would YOU?" she pleaded with an encouraging smile and praying hands.

"Yeah, I don't mind." He replied pulling the car on over and hopping out.

He walked to the rear of his car, popped the trunk, got his jack and lug wrench out, and slammed the trunk heading to Mrs. Owens' car.   As his dad got started with the tire change, Antwan sat on the front doorstep with Mrs. Owens and fidgeted with his smart phone. Mrs. Owens marveled at his dad thinking of what could be. He seemed to have the looks and a nice heart too. He had a nicely trimmed fade haircut and beard, looked to be around her age, 45 or so, and had pretty white teeth and nice chest muscles and biceps. She began thinking, *"Is it selfish for me to desire this man to be MINE and not work things out with his wife?"*

   He asked her where her spare was, and she answered. He proceeded to the rear of her car and retrieved her spare out of her trunk while Mrs. Owens studied his rear in his shorts.

"You just don't know how much I appreciate this." She remarked.

"Oh, I don't mind giving a lady some help." Luther replied.

"I'm gonna have to invite you to dinner or something NOW!" she added.

"Oh, now you punching the right buttons!" he laughed. "I love a good meal!"

"I'll fry some chicken, make some mac and cheese, cook some collards with smoked hamhocks, and bake a sweet potato pie. How's THAT sound?"

"Oh, THAT sounds DE-LISH-SHUS! You got my mouth waterin!" Luther commented.

"Well, you're on! How's Sunday sound?"

"What time?" he snickered as he continued working.

"Ohhhhhh, Let's say 3:00!"

"Sounds like a winner!" Luther replied.

A fifteen minute silence followed and then Luther was done. He got up brushing his hands together and smiling showing those bleach white teeth.

"That'll be a home cooked dinner with plenty of sweet tea, please!" he concluded with a bright smile.

"Coming right up on Sunday at 3:00 PM! THANK YOU SO MUCH!" Mrs. Owens grinned as she stood up extending her hand out to him to shake his hand.

"My hands are all dirty. Let's do a fist bump." he told her.

Luther and Mrs. Owens bumped fists and exchanged smiles. Then he and Antwan headed for his car and within minutes were heading on down the road.

Mrs. Owens had a good feeling about Mr. Pendergrass. He was gallant, heroic, good-looking, AND kind on top of it. Yes, Mr. Muscles looked good all right, but he was a real jerk when it came to doing something other than what he wanted to do. Plus he wasn't a REAL man that could stick around longer than twenty minutes without a person having to mash a damn button to bring him back. She felt kind of sleazy on giving it up to Mr. Muscles looking back, but she had gotten so horny it was a shame. So, she was going to try and move past that little rendezvous in hopes for bigger and better more real things.

***

16

Lance Fielding sat in his office thinking. He had taken LaShonda's dismay about the use of her husband's image in the app to the board and had gotten another $2,500 per month royalty approved. She was already getting $1,800 per month just for Dexter's share in the research for the app and for the app bearing their last name along with his in its name: The Fielding-Smith App. He snatched up the phone and delivered the news.

"Hello?"

"Hello."

"LaShonda. This is Lance. "

"Yes?"

"I was calling to tell you that I got you another $2,500 royalty approved on this app business."

"Oh?"

"Yeah! It'll be starting next month after it goes through Accounts Payable in our Accounting Department."

"Well, thank you. That's a step in the right direction. I still think I should have been the first one contacted about using my husband's image like that. Well, I have things to do, but thanks again."

"You're welcome. Take care."

He sat looking at his family pictures on his desk. He really adored his wife, Lisa; his daughter, Loren; and his son, Peter. He smiled rubbing his finger across the photos as memories of each crossed his mind. They were the reason he got up and went to work every day. The perfect motivation! Even though Loren was in her freshman year at UCLA, she was still costing him money—money he didn't mind spending since it was for a good cause.

***

17

Loren Fielding dropped some clothes in her washer and started it. She then decided that she'd go jogging while they washed and return to transfer them to the dryer when she got back. She grabbed her cell, her earphones, and a cold water bottle out her fridge. Then she plugged the phones in to her cell and placed them on her ears and headed out her apartment door. She did some squats to warm up and then started running. Her brunette hair bounced on her back in a ponytail. She strolled by the park as she enjoyed the scenery. Her classes were going well and her social life was doing okay also. She felt at one with nature, a greatness and ambiance she couldn't describe. She dashed by some woods on the other side of the park as she listened to Adele singing "Hello" and enjoyed the soothing sunrays beaming down from the afternoon sky. Traffic was sporadic. She only saw a few other students either walking or jogging along the way. She ran on ahead to the campus library. It was there that she decided to U-turn and head back to her apartment.

   She made an about face and headed back toward home. She was now covered in sweat, but it felt good. She knew that sweating helped get the poisons out of one's system, so she welcomed exercise. In ten minutes she was besides the thick wooded area again. This time, some a guy's arm reached from the bushes and snatched her into them. Before she could scream, whoever it was had placed one hand over her mouth and had his other arm around her back. He snatched the earphone from her ear and dared her to scream with his hand over her mouth. He placed a leg behind her and pushed her backward forcing her to collapse on the grass and leaves beneath her. Upon her fall with him falling on top of her, she felt her phone try to topple out of her pocket, but she managed to tuck it back in just in time. The guy had a scarf around his face but she knew he was white by his arms and hands. He began to undo her clothes in a frenzy. She thought about her dad's new app on her phone. She slipped her hand back in the pocket to her cell to depress the "H" button. In a second's time there was thick white smoke which cleared a second later. A force snatched the masked guy off of her. Loren was astonished. She jumped up to witness what was happening.

Some muscular black guy was laying into the guy with the scarf. She thought to herself, "This must be the Dexter Smith guy who her dad used to work with that they had told her about." Her mom had told her about him, but she hadn't paid her too much attention other than to learn to depress the special button when she needed help. The black dude was all muscled up with perfect biceps, triceps, and pecs in his chest and only had yellow Speedos on with socks and combat boots. He was punching her attacker so violently that she felt like making him stop for a brief moment just before he stopped by his own volition. He pulled her attacker's hands together and slapped handcuffs on him that magically appeared into his hands from the thin air. Finally he gazed at Loren to speak to her.

"Young lady, the cops are on the way. This piece of shit won't be attacking anyone else! You're one of the lucky ones! Your dad's new phone app came to your rescue, BIG TIME!"

He sat down on the turf beside the white dude and smacked him side the head again. Then, he held out his hand, snapped his fingers, and a large fruit flavored Powerade bottle magically appeared out of clear blue sky. He unscrewed the cap and started gulping it down. He had worked up quite a thirst with the beat down he had given the attacker. Loren sat on a log facing the two of them as they all waited for the police. She couldn't believe this was actually occurring.

Within a few moments, the cops could be heard coming with a siren blaring. Dexter Smith leaped up and dashed out the clearing to the roadside and started waving his arms to get the attention of the patrol car. The car stopped and a cop jumped out. Dexter led him to the embankment inside the woods where the attacker sat handcuffed facing Loren.

"Okay. What do we have here?" the cop asked.

"Officer, I was jogging and when I got by these woods, this guy sitting here tried to grab me and rape me, but I activated my Self Defense App on my phone, and this guy appeared and came to my rescue." Loren explained.

"I see. So, are you alright?" he asked.

"Yes, thanks to THAT guy, he didn't have a chance to harm me." Loren said pointing to Dexter.

"What's your name, mam?"

"Loren Fielding."

"And you sir?" he said to Dexter.

"Oh, I'm just an illusion. I can be seen and felt, but I'm really a figment of your imagination. That phone app her daddy and I invented is what brings me back for people's assistance."

"You say, you and her daddy invented the app?"

"Yeah, I am the virtual reality illusion of my former self, Dexter Smith, and I used to work at Galactic Corporation with HER dad, Lance Fielding, before my untimely death."

"Ohhhhhhh, I see."

"Well, thanks, Sir, for your help! I'll take it from here!" the cop said.

He reached down at the attacker and snatched him up by his forearms.

"Come downtown and pick you up a copy of the report, Mrs. Loren. It should be ready by tomorrow afternoon. Let's go, Mister. You have the right to remain silent. Anything you say will be held against you. You have the right to an……"

The cop's reading of the guys rights faded as they cleared the opening of the woods to the awaiting patrol car on the street side. Now it was just Loren and Dexter's illusion alone in the clearing. He walked over to her. His perfect body was sweaty and oily, but sexy as hell. She got a good eyeful of his sultry chocolate body as he approached her. Lusty thoughts crossed her mind, and she fought them off desperately. He began to speak.

"You take care of yourself, okay?" he said.

"Okay," she answered. She felt like snatching his Speedos off and getting busy with him. She gulped trying to swallow down her unladylike desires.

"And keep that phone within reach at all times. It really saved you today. Tell your dad I said, 'Hello'. It's about time for me to disappear, you know?"

"Disappear?" Loren questioned.

"Yeah, after I complete a rescue, I automatically vanish shortly afterward."

"Okay. Thanks for saving me!"

"Anytime," he finished with a smile.

No sooner than he got it out of his mouth, he vanished leaving her alone standing there. She reached into her pocket to feel for her phone. Once she reassured herself that it was in place, she began jogging back toward home to safety. She had a burning desire to call her dad and tell him how the new app he and his friend created had saved her. Plus Dexter, her dad's late friend, had her craving the chocolate ice cream in her freezer like never before.

***

18

Simon McCorkle drove his classic Mustang up to the gas pump on the west side of town. His needle wasn't working so he didn't really know how much gas was in the tank at any particular time. He just had to fill it all the way up whenever he could and estimate how much was in the tank by how many miles he had driven on his odometer. But, apparently, he had a gas leak also since he smelled the gas fumes quite loudly. He was going to have to get it to the shop ASAP, but right now all he knew to do is get it filled again. He zipped out his debit card and inserted it into the machine. An SUV zoomed up next to his car, and a guy jumped out the back pointing a pistol at him. He dashed to Simon as he continued pointing the pistol.

"Motherfucker, where the keys at?" the black dude demanded.

"Right here in my pocket," Simon answered.

"Well, pull 'em out, pimpin….Befo I go up side yo head with this damn gun!" he taunted.

Simon reached into his pocket for the keys. He was so stressed that he forgot about his cell phone being in there too. He hurried and felt for the caged 'H' button and depressed it as quick as possible before he snatched the car keys out. Before he could hand the key over, white smoke appeared mysteriously just as Simon had witnessed in the board room at work. The smoke cleared and left the muscled up black hero he'd heard so much about, none other than his old work mate, Dexter Smith who angrily pointed a 9 millimeter pistol in his right hand to the culprit. Dexter got the keys from Simon's hand with his left hand and then jingled them in his fingers with an angry look.

"Are THESE what you're wanting, punk?"

The guy stood with his mouth stretched with wonderment on the things he was witnessing.

"Get your black ass away from here befo I kill ya!"

The guy dashed back over to the open door of the SUV and jumped in. Then, the SUV sped off. Simon stood speechless looking at Dexter.

"Thanks, Dex!" he hollered.

"Simon, how have you been man?!"

"Fine, until today! This guy has my heart about to jump out of my chest, man!"

"Thank God for this app thing!"

"Yeah, it paid off, huh?"

"THAT it did!"

"Lucky, I had my phone in that very pocket I had my keys in or I would have been without my car!"

"Yeah, good thing it WAS in there with the keys."

"Wow! This is absolutely incredible! We miss you man!"

"Yeah, I miss you all too! I'll be disappearing in a moment. Take....."

Dexter vanished before he could finish his goodbye.

Simon mashed the button again to let him finish his greeting. He felt so sorry for him being in this sort of predicament out of his control. Dexter was back momentarily once the smoke cleared. They began chatting again while Simon completed getting his gas. Then, he asked Dexter to get in once he got ready to drive. They rode talking about this and that until Dexter vanished again. When he vanished again, they were at a good pause and had gotten to a nice point in their conversation, so it didn't really warrant a third press of the button to bring Dexter back. Simon kept on driving and missing his old friend all the while. He couldn't wait to tell Lance about the fiasco when they got back to the office.

***

19

Lashonda pulled out onto the street headed to school. She started her mornings every weekday at 7 AM. She'd get up, get dressed, grab a microwaved biscuit, and head out the door to get to her class by 8 A.M. She loved teaching. After Dexter's demise, she poured all of her energies into her teaching even moreso. She got to the highway and made her right turn headed downtown. She drove on thinking about this and that. She thought about Dexter being trapped in his own world, and the claims of infidelity. Then, she realized that he wasn't real- that he was really dead. He was alone just as she was….alone. She couldn't really be mad at him. How COULD she be mad at him? What kind of life was that? To be trapped and knowing you had a wife and kids that you could only see once you were summoned back by a button. That wasn't a good life. And then to have horny women taking advantage of the chance to be with the sexiest man that ever lived, in Lashonda's point of view. Her Volvo crossed an intersection and everything went black.

***

20

Lance sat once again in his office. He couldn't believe that both Dexter and LaShonda were now together once again. He thought about how fragile and unpredictable life was. He thought about how dangerous it was to drive on our highways. LaShonda had been broadsided by a utility truck that had run the red light. She didn't even know what hit her. Lance realized that people feared flying on airplanes, but statistically they were hundreds of times safer than cars ever dared to be.

Lashonda's funeral had been especially sad. The two youngsters were in there cute little black outfits, but they had a sense of what was going on it seemed judging by their irritability. Her parents took it hard, but knew that it was up to them and their other two daughters to raise the kids. They had to be strong for the kids' sakes.

    Lance shuffled his way through the day, and before he knew it he was in his lab out behind his house. He pulled out a picture of LaShonda and studied it. Then he scanned it into his computer and sat looking at the computer image on the screen. Then, he downloaded the image into his hieroglyphic quantanomizer and let the machine crank numbers and spit out its results. After ten minutes, the green light atop of the machine lit up. Lance smiled. He then pressed "Display". The lights in the room dimmed and beams of light began to swoosh around the room. Then the beams aimed all in one spot providing various colors to the spot. Then some white smoke appeared and cleared out very quickly. Standing in the room was LaShonda's hieroglyphic likeness. All Lance had programmed into the machine for her attire was a swimsuit and sandals, so that's all she had on. She looked at Lance and snatched herself around in a circle wondering what was going on.

  "Lance?" she quizzed

  "Yes, LaShonda?" He answered.

  "What happened? Why am I here? Where are my kids?"

"LaShonda, you were in a tragic accident. You are now a hieroglyphic like your husband. You two can be together now….forever."

"Oh my GOD!" she shrieked.

Lance mashed his cell phone 'H' button and Dexter appeared after white smoke appeared and cleared. Dexter stood facing LaShonda.

"Dexter?"

"Baby, it's YOU! How can this be?"

"Lance said I was in a tragic accident, and now I'm really dead, but he has me as a hieroglyphic like you."

"Now we can be together, BABY!"

"That's good in one way, but now my kids….OUR kids are orphans!"

"I realize that, but your sisters and your parents will take good care of them. I'm not worried about THAT! Yes, it pains me that I can't have a normal relationship with them anymore, but Lance could have your parents or your sisters mash the button so we could see them."

"But we'll be trapped in here looking like we do now FOREVER while our kids grow up and my parents grow old and die and my sisters get to be old people." LaShonda explained over falling tears.

"I guess I didn't think THAT far ahead." Dexter muttered with a tear of his own dripping from his eye.

"Listen, guys, I thought of something." Lance butted in with watering eyes of his own.

"I'm going to let you two be free once again. This was a grave mistake bringing you back only to witness what you can't enjoy as normal living adults anymore. Whenever you give me the cue, I'm going to undo this. We could get use genderless nonfacial hieroglyphic likenesses for this app just as well. It would do the same job, but would tie you two or anybody else of humankind into this realm of servitude and inescape."

LaShonda looked into Dexter's eyes. He grabbed her by the hands and kissed her cupped hands.

"Okay, take us both to see my kids and my parents. Then, let us go back on our way to the other side."

"Okay." Said Lance.

He downloaded his computer program in the hieroglyphic machine into a USB drive and then plugged the drive into his phone. Now, he had both LaShonda and Dexter in the app.

Lance told Lisa a quick rundown on what was going on, and she had Peter stay at his friend's house while she rode with Lance over to LaShonda's parents' house on the other side of Seattle. They knocked at the door, and her dad opened the door.

"May I help you?" he asked.

"Yes, are you Lashonda Smith's dad?"

"I am, and who are you?" he asked with peaked curiosity.

"I'm Lance Fielding.....the man who created the phone app with your son-in-law in it?"

"Ohhhhhhh, YOU'RE the one! Come in!"

Lance and Lisa made their way to the living room and stood there awkwardly looking at him and his wife.

"Well, have a seat!" they suggested.

They sat down on the loveseat. It was a nice and cozy inside their house.

"I came here on a mission, sir and mam, to show you something." Lance stated.

"To show us what?"

"To show you this." Lance said as he pressed the 'H' button on his phone.

Smoke came and cleared and left both LaShonda and Dexter standing before them. They both leaped up. "Oh my God! Dexter…..LaShonda…..? Where are yall's clothes at?" Her mom dashed to the linen closet and got two big towels. "Here, cover up!" she said handing both of them the towels.

"Mom, I want to peep at the kids." Lashonda asked.

"Go head baby!" her mom said.

LaShonda and Dexter strutted to the bedroom with the towels wrapped around them. The kids were sound asleep. They kissed them on the cheek and left out the room. They came back to the living room and sat down. They began talking amongst one another.

"Mom, we're contemplating going on to the other side for good. We don't want to be trapped in this app business forever."

"But we're just now seeing YOU again, LaShonda!"

"Yes, but this isn't right."

"Is there not a way that we can see you like this and you NOT be in everyone's phone across the nation?"

"There is," Lance interrupted. Lisa stared into his mouth waiting for his reply.

"How?" both Dexter and LaShonda inquired with eager anticipation.

"I can deprogram the app server, and all across the country it won't work anymore. Then, I could reprogram it with a neutral gender neutral image like an avatar or something, and program it so that only one or two phones could bring you two back. LaShonda, your mom and dad, could bring you back as they wish or however you work it out using these two particular phones like a his and hers. How's THAT sound?"

"Sounds good!" they said looking at one another.

"I'll be happy doing it only until the kids get to be adults, I think. That'll be time enough. Then, it'll be time to part with the idea teetotally." LaShonda said looking at Dexter who nodded his approval.

"Well, I'll jump on it as soon as I get back to the house." Lance stated.

Both of LaShonda's parents hugged her one at a time. They couldn't believe that they were having a chance to hug her once more. Everyone said their "Goodbyes" and as Lance and Lisa walked out the door, Dexter and LaShonda vanished. Everyone looked at one another and smiled. The Fieldings loaded up and left.

***

21

Six months went by. News began to get out about a new inhumane hero in a soft silver metallic suit. This hero had a face like the statuette handed out during The Oscars. He didn't have the sex appeal as the former "Mr. Muscles" dude, Lance's friend, Dexter." Sales dipped significantly for these months. People were tweeting comments like "Bring back Mr. Sexy!", "It's just not the same!", "The new rescuer is ugly!" and "What happened?....the black dude dodging the child support he done racked up?"

Dexter and LaShonda had fun being brought back by her parents to play with their toddlers on weekends. They'd meet up for an hour or so and play with them until bedtime. Her parents would keep pressing the magic button over and over to keep them around for the required time. Once the agreed time arrived to part with their kids, the disappearing act would not be reversed. Then, once they disappeared, her parents would clean the kids up and put them to bed. And this would do until the NEXT weekend came around……..And the next weekend….And the next. Together, they liked it better this way. They planned to keep up this routine as long as at least one of her parents was around, and for the youngsters to eventually be shown how to do the summoning once they were old enough to grasp the button mashing concept.

Gone were the days of Dexter popping up all across the nation in people's houses, in parking lots, and in marketplaces just to rescue them or scold them to get their babies out of the baby seats. And gone also were the bogus rescue calls of women just anxious to see the hunky chocolate muscleman.

***

22

Latoya sat on the bar stool sipping on some Alize. She could hardly wait to get the party started. Her girl, Peggy Owens, had told her all about her rendezvous with Mr. Muscles. Penny was taking a little longer than expected returning to the living room.

"Girlllllll, was takin' you so long?" Latoya hollered down the hall after Penny as she ran her fingers through her frizzy red hair.

"Okay…Okay….Here I am!" Penny responded as she entered the living room. She walked over to the Alize and poured her up some in her champagne glass. She picked it up, took a sip, and then walked over to coffee table. She pressed the "H" button on her phone, and instantly white smoke was everywhere just like they had been hearing about. Then, the smoke cleared within seconds as the rumor suggested leaving a Martian looking manly statue in front of them. It was six foot tall and silver from head to toe.

"Is some...one in dang...er?" it asked.

"What the hell?" Latoya shouted in disgust.

"What kind of shit is this?" Penny followed up in fury.

"I'll be damned! I wanted to see this hot....sexy....chocolate...muscle man I been hearin' so much about! Not this....this....man sized Oscar trophy! Where Mr. Muscles at?!" Latoya angrily vented.

"If –you- don't- need- my- assist-ance, just-mash-the-butt-on-again-to-send-me-back." the strange silver man suggested in a robotic tone.

Penny eagerly depressed the button again and the strange silver guy vanished. They both sat fussing amongst each other until Latoya's phone rang.

"Hey, where you at?" her boyfriend, Keandre, questioned.

"Oh, I dropped by my girl Penny's house for a minute, and now I'm fixin' to head home." Latoya responded.

"Well, I'm waitin' on you. I'm ready to eat! Hurry up!"

"Okay. See ya in ten minutes."

She ended the call and looked around at Penny.

"Girl, I thought we was fixin' to see sump'm. But um bout to go. Keandre waitin' on me."

"Okay, girl" Penny responded.

Latoya strutted on out the door with her purse.

Penny sighed and sipped some more Alize. Her phone rang.

"Hello." she answered.

"Hey, what's goin' on?", her boyfriend Johnathon interrogated.

Her eyes lit up and she sat back on the sofa.

"Hey, Johnathon. Are you comin' by?"

"Sure."

"I'm hungry. Can you bring some Chinese?"

"You got it. I'll bring some shrimp fried rice."

"Sounds GOOODD! See ya shortly then" she smiled.

She sat on back on the sofa finishing up her Alize with a smile on her face. "I already GOT me a Mr. Muscles anyhow!" she said in a low voice as she did a little wiggle.

THE END

Thank You!

I thank you very much for purchasing this book. I hope you enjoyed it. I invite you to write a short book review on it to give me your feedback and post it on my Facebook page or inbox it to me. You'll be pleased to know that I am currently working on other books of various topics that I look forward to having you read and enjoy as well.

About The Author

Kevin Brady is the eldest of five kids and a graduate of Statesboro High School (1985) and Georgia Southern University with a B.B.A. in Business Management (1998). In 2015, he published three other books including *Ghostly Encounters*, *The Sinless House (the 1st book he and his wife, Charlene, teamed up on)*, and *A Helluva Home* . In 2016, he published *Fat-Man: the Dream Hero* and his *K.D.Brady's Four-In-One Holiday Collection,* and in 2017, he published *Funny Things Mama Said and Did, Won't God Do It?, Blood Thirst,* and this book, *The Officer.* He is married to his wonderful wife of eleven years, Charlene Brady, and has one stepson, a daughter-in-law, and three adorable grandkids. He loves old car shows, ghost documentaries, reading, writing, going to church, fishing, playing lottery, golf, watching championship sports, barbecuing, and being with family and friends. He is working on other stories that will be available as soon as they are ready.

Contact Info:

Facebook: https://www.facebook.com/#!/kevin.d.brady.9

Email: kdbrady6798@yahoo.com

Book Links:

Ghostly Encounters:

https://www.createspace.com/5612057

The Sinless House:

https://www.createspace.com/5696025

A Helluva Home:

https://www.createspace.com/5824467

Fat-Man: the Dream Hero:

https://www.createspace.com/6194720

K. D. Brady's Four-In-One Holiday Collection:

https://www.createspace.com/6612177

Funny Things Mama Said and Did:

https://www.createspace.com/6958606

Won't God Do It?:

http://www.createspace.com/7057157

Blood Thirst

https://www.createspace.com/7297479

The Officer

https://www.createspace.com/7504472

The Self-Defense App

https://www.createspace.com/7824317

Website:

http://kdbrady6798.wixsite.com/mysite

www.ingramcontent.com/pod-product-compliance
Lightning Source LLC
Chambersburg PA
CBHW020436220526
45464CB00002B/723